John Abercrombie

The Garden Mushroom : Its Nature and Cultivation

A Treatise Exhibiting Full and Plain Directions, for Producing this...

John Abercrombie

The Garden Mushroom : Its Nature and Cultivation
A Treatise Exhibiting Full and Plain Directions, for Producing this...

ISBN/EAN: 9783337009533

Printed in Europe, USA, Canada, Australia, Japan

Cover: Foto ©berggeist007 / pixelio.de

More available books at **www.hansebooks.com**

THE
GARDEN MUSHROOM;

ITS

NATURE AND CULTIVATION.

A

TREATISE,

EXHIBITING

Full and plain Directions, for producing this defireable Plant
in Perfection and Plenty, according to the true fuccefsful
Practice of the LONDON GARDENERS.

BY

JOHN ABERCROMBIE,

Author of MAWE's GARDENER's KALENDAR.

LONDON,

Printed for LOCKYER DAVIS, in HOLBORN.
M DCC LXXIX.

Price One Shilling and Six-pence.

ADVERTISEMENT.

IN the courfe of forty years practice and obfervation, I have generally remarked, that the culture of the Garden Mufhroom has proved confiderably more precarious and unfuccefsful than that of any other kitchen-garden

A 2   vegetable ;

vegetable; or even of almoſt any other cultivated plant of our gardens; and that its true nature is little known among the generality of gardners. Some, even amongſt experienced and ingenious profeſſors, who raiſe all other plants in perfection, have been found often to fail in the article here deſcribed.

This plant is of ſo very ſingular a growth and temperature, that, unleſs a proper idea of its nature and habit is attained, and the peculiar

culiar methods and precautions
purfued in the procefs of its propagation and culture, little fuccefs
will enfue. The whole management of it remarkably differs from
that of every other fpecies of the
vegetable kingdom ; and it is
the moft liable of any to fail,
without a very ftrict obfervance and
care in the different ftages of its
cultivation.

Directions refpecting the culture
of Mufhrooms, are to be met with
in various books of gardening,
but

but they are defective and prove
to be of fmall fervice, becaufe they
are not the refult of real practical
experience.

CONTENTS.

Spawning

CONTENTS.

Of

Of the UTILITY of the

GARDEN MUSHROOM,

AND

Its Preference to the FIELD SORT.

———————————

THE *Garden Mushroom*, or that produced
by the art of horticulture, or procefs of regular cultivation in gardens, is greatly fuperior in all refpects to the wild chance
Mufhroom of the meadows and paftures. It
is now univerfally admired as one of the de-
licacies of the kitchen garden; and is a requifite production of that department; being
always in requeft, and highly acceptable,
though feldom obtained in plenty and perfection. This difficulty has been owing to
its fingular mode of culture being little

B   known

known to the far greater number of practi-
tioners beyond the vicinity of London.
Here it is raifed by many of the kitchen-
gardeners. Their fuccefsful method we
have long practifed, and now propofe to
explain to the reader.

It may be juft obferved, that although
the Mufhroom grows fpontaneoufly in meadows and pafture-fields, it is obtained there
accidentally only, and at a particular feafon :
but, by garden culture, we procure this
plant at any time of the year, whenever it is
wanted, and always of fuperior goodnefs,
richnefs of flavour; and with a certainty of
its being the true falutiferous, or wholefome
kind : a matter of the utmoft moment, fince
there are, in the fields, fo many of a perni-
cious quality, bearing fo great a fimilarity
to the true fort, that, having been gathered by
the unfkilful, have proved fatal to thoufands,

Nature,

Nature, Mode of Growth, and Specific Dif-
tinction of the wholefome Species from the per-
nicious Kinds.

THE Mufhroom is a fungous plant,
without appearance of leaves, flowers, or feed,
a fpecies of the *Genus Agaricus,* fubject to the
botanic clafs *Cryptogamia Fungi,* comprehen-
ding fungous plants, which have concealed,
or doubtful genital organs, and without vifible
flower or feed; thereby belonging .to the
family of imperfect vegetables; a numerous
train of which are of this fungous tribe. They
confift of different *genera,* and numerous fpecies and varieties, many of them of a poifon-
ous, or at leaft of a fufpicious nature. One
fpecies only merits cultivation as a wholefome
efculent, which is that under confideration.

Agaricus Campeftris, Field Agaric, or Com-
mon Mufhroom, rifes from the ground in its

perfect

perfect form, with an erect robuft *ftipes* or ftem, one inch or more high, crowned with a round, convex, thick, flefhy, white head or hat, *pileus*, with *lamellæ*, or gills, underneath, of a reddifh flefh colour; it is fuppofed the flower and feed, if any, are concealed between the *lamellæ*. When the plant arrives at full growth, the head expanding almoft flat, forms a large flap, and falls on the ground over the fuppofed feed.

This is a fugacious plant of quick growth and fhort duration, advancing firft like fmall white round knobs, which, increafing faft in fize, and fometimes partly accomplifhing their growth within the furface, fuddenly make their eruption from the earth above halfgrown, in the morning, where there was no fign of them the night before. But in the regular beds they arife varioufly, fpreading over the whole furface, fome as fmall as peafe,

peafe, fome the fize of buttons, and fome
near full growth; others frequently appear
iffuing from the bed of a large fize, being
completely formed under the furface.

This fpecies (*Agaricus Campeftris*) is diftinguifhable from all others, by its fine white
flefhy head, the red colour of the gills, and
by its imparting an agreeable Mufhroom
flavour. As the plant becomes large, the
gills affume a blackifh red without, retaining
however internally its flefhy colour, by which
it always fhews itfelf to be the true fort.

Generation of the Plant.

THE Mufhroom tribe has long afforded
much fpeculation to naturalifts, with refpect
to being perfect or imperfect plants. The
flower and feed, from their exceeding mi-
nutenefs

nuteness and obfcurity, (if they really exift at all) remaining invifible even by the aid of the microfcope. Many therefore fuppofe that there is no production of flower or feed, but that they owe their origin entirely to the putrefaction of earth or dung. This fort of foil however firft difcovers them under the form of a white, mouldy, fibrous fubftance, called fpawn, which proves productive of nume-rous minute white knots, or embryo plants, gradually increafing to the perfect Mufhroom.

On this fubject the botanic world have been long divided in opinion, the moderns generally contend for the flower and feed, and have in a manner confuted the doctrine of putrefaction.

Thefe confider the Mufhroom as a true and perfect plant, produced from feed afforded

afforded from the flower of preceding plants by fome wonderful fecret in nature.

The invifible feed difcharged on the adja-cent foil, and thence diffeminated by the air to fituations adapted to it's nature, germi-nates and fhoots forth into white fibrous, cobweb-like fubftances, fpreading and form-ing the fpawn and embryo plants for the production of the future Mufhroom. Thus, probably, by fuch diffemination, and fo myfterious a progrefs of nature, adapted to certain foils and fituations, it is, that we often find both Mufhroom and fpawn abundantly in obfcure places where none were ever obferved before, in old dung hot-beds, horfe dung-hills, and in bye dry places where horfe-ftable dung has lain undifturbed till rotten.

The fpawn is alfo often found in pafture fields under the turf, in places where Mufh-

rooms

rooms are obferved to rife naturally, dung-
fpawn however is preferable for garden culture, as well on account of the probability of
a good crop, as of it's being the true fort.

Of collecting the Spawn.

AFTER the foregoing ftrictures on the
general nature, growth, generation and propagation of the Mufhroom, we proceed to explain the neceffary preparation for its culture
in gardens, by the following directions concerning the fpawn.

The propagation of the Mufhroom is to
be effected by planting lumps of fpawny-
dung, found chiefly in dry rotten dung or
clods of dungy earth, and interwoven in the
foil in numerous white ftringy fibres, often of
a cob-web-like form, and if of the true fort,
disco-

Fig. 2

discovering a strong smell of the Mushroom. A due quantity should always be provided previous to making the bed, in order that you may more readily judge of what size to determine upon; for it is sometimes difficult to be had in any considerable abundance; so that according as it is occasionally met with, it should be carefully collected, taking the lumps of spawn and earth entire, of which, for a bed twenty feet long, three or four bushels will be requisite, and so in proportion.

Spawn is obtained the most readily and in abundance in parcels of decayed dung and dungy composts; but commonly more plenteously and good in rotten horse stable dung, composed of the short dung and moist litter together, as cleared from the stables, either collected in dung-heaps, or formed into hot-beds, composts, &c. when it has

C   remained

remained some months till its fermentation and heat are decreafed, and a ftate of decay and putrefaction brought on. This kind of dung being more adapted to the generation of fpawn than any other, is a favourable circumftance, as horfe dung is to be every where met with.

In cucumber and melon beds, at the end of the Summer, when the crops are over and the dung decayed or rotted, we often difcover great plenty of moft excellent fpawn.

Old Mufhroom beds likewife which have been compofed of the fame kind of warm dung, when decayed or worn out, and pulled to pieces, generally afford good fpawn, which fhould be carefully preferved till wanted.

Re

Be careful too in searching adjacent old dung-hills and dungy compost heaps in any out-grounds, and in stable yards, where horse dung-heaps have been for some continuance in a state of decay, especially in obscure dry corners long undisturbed.

In the horse rides and livery stable yards in and about London, where the long covered rides are littered thickly from the stable, with occasional dunging and staling, search towards the sides where you will often find great abundance of fine large cakes of most excellent strong spawn.

Horse-mill tracks also, where horses are constantly employed under cover in turning mills; and many of the great London breweries, tan-yards, and large manufactories,

C 2

where

where horfes work under cover, frequently furnifh very fine fpawn. .

Sometimes in kitchen gardens, when the ground has been thickly dunged in the Spring with half rotten dung, on digging the fame again in Autumn, and looking with care, good lumps of fpawn are to be found.

So that in all decayed dung-heaps and hot-beds, old dungy compofts, and well dunged foils, not too wet, or the dung very buttery rotten, you may be fuccefsful.

Frequently in old dung-heaps fome ftragling Mufhrooms are feen to rife naturally in Autumn, there you may be fure to find fpawn.

Laftly

Laſtly it may be procured in the meadows and other graſs paſtures towards the end of Summer or in Autumn, here and there, in places where Muſhrooms happen to riſe in their natural growth : breaking up the turf, the ſpawn will be found in the earth, and may be digged up in lumps for uſe : however, where enough of dung ſpawn can be had, I always prefer it to that of the field, as before noticed,

The beſt ſeaſon to find ſpawn in the greateſt plenty and perfection is the Autumn and early part of Winter; for ſpawn being of a ſingular temperament, impatient of much wet, or cold, or of being much expoſed to the open air, it ſhould be carefully collected for uſe before it is injured and weakened by the inclemency of the weather ; for it is of much importance to have it in full vigour, when, it may be directly uſed

in

in fpawning beds, provided it be quite dry; otherwife let it lye by for a few weeks.

Be careful, in collecting the fpawn, to have the lumps or cakes of fpawny dung taken up entire, placing them in a bafket or wheelbarrow, in order to be carried into fome dry clofe fhed or room, to be depofited till wanted; noticing whether any of the lumps be wet: in that cafe fpread them to dry a little; then let the whole be placed in a dry corner, clofely covered with ftraw or litter or garden mats; or packed up in facks or hampers, covered clofe in the fame manner, whether for prefent ufe, or for keeping. By attending to thefe directions its vegetative power may be long retained, and the fpawn fafely fent to any diftant place.

We fhould be particularly cautious to reject fpurious or falfe fpawn; for there is a

degenerated

degenerated variety, called white-cup, which
produces a fort of Mufhroom with a fmall
thin white head without any flefhy part, and
generally rifes up fuddenly in the beds. This
fort is entirely ufelefs, and often difappoints
the gardener. It is diftinguifhable generally
by its great abundance all over the lumps, by
its very fine filky cobwebby nature, and its exceeding white hoary-like appearance; it has
little or no fubftance, and emits but a very
faint fmell of the Mufhroom.

About London, where great quantities of
Mufhrooms are raifed for the markets, and
confequently vaft fupplies of fpawn are an-
nually required, there are experienced Mufh-
room-men, who, at the proper feafon, go
about collecting, both in town and country,
the true fort, which they buy commonly from
about half a crown to five or fix fhillings per
bufhel, according to its goodnefs or plenty.

In

In very cold wet feafons it is both bad and fcarce; and dear in proportion.

Good fpawn may alfo be purchafed occafi‑onally of the kitchen-gardeners in the neigh‑bourhood of London, many of whom have extenfive Mufhroom-beds, as well as com‑mon hot-beds. Thefe beds when old, being pulled to pieces, often afford more fpawn than the gardener has occafion for, which they lay up dry, and difpofe of by the bufhel when wanted.

Let it be obferved again of the fpawn in general, that it muft be kept dry till wanted; and if any lumps at firft gathering appear wet, fpread them in a fhady covered place before they are laid up in a houfe; for it is of much importance to have the fpawn perfectly dry when planted.

Of preparing Dung for the Beds.

NO dung anſwers the purpoſe ſo well as that of the horſe, the dung and urine of this animal, together with the wet ſtraw litter of the ſtalls in the ſtables, being of a hot quality, ferments, and acquires a ſtrong degree of heat of long duration; but as this heat generally proves too violent at firſt for the growth of vegetables, the dung ſhould always be previouſly reduced to a proper temperature, by caſting it up in an heap, and turning it once or twice, in order to evaporate the rank burning ſteam before its fermentation. A quantity, in proportion to the ſize or extent of the intended bed, muſt be procured. For a bed of twenty feet long, three or four large cart-loads will be neceſſary, and ſo in proportion to any length intended; as a bed may be made of almoſt any extent,

D   from

from ten feet to fifty if required ; four or five feet wide at bottom, drawing into a sharp ridge at top four or five feet high ; which will allow for settling.

For private use, a single bed of about ten or fifteen feet in length may be fully sufficient. But for the supply of the London markets, long parallel ranges are made, from twenty to fifty feet in length.

Provide therefore a proportionate quantity of the best fresh horse-stable dung and litter, warm and moist, rejecting such as is dry and decayed, and such as has already exhausted its fermenting property. Let this be taken long and short as it comes to hand; and as it is brought in, toss it up together in an heap, carefully mixing, that the whole mass may acquire an equal degree of heat.

Thus

Thus let it remain together three or four
weeks, according to the quantity and ftrength
of heat, in order that it may meliorate, by
difcharging the rank obnoxious fteam; and
if it is turned over once every week, it will ftill
incorporate the parts more effectually, and
give an additional vent to the fierce ferment.

This preparation of the dung is abfolutely
neceffary, as without fuch precaution, when
formed into a clofe bed, it is apt to acquire
fuch a vehement degree of heat, as to burn
and exhauft its vegetative power, without
being able to effect the purpofe intended; for
the fpawn requires a bed that only gradually
advances to its full heat, and declines in the
fame gradual manner; till reduced to the low,
kindly, growing warmth that is peculiar to the
nature of the fpawn, and the growth of the
Mufhroom:

D 2

Of

Of the Mushroom Bed.

THE seasons for making the beds have been already observed under the article **of** preparing the dung. With respect to the situation, they may either be in melonary or cucumber **ground, in a** dry elevated spot, and a warm sunny exposure; or in any **of the large** quarters of a dry kitchen garden.

They **may be** made either **entirely on the** surface, or occasionally in a shallow trench. In low or strong **soils,** where there is danger of water remaining in Winter, or after hard rains, elevate the bottom of the bed sufficiently from the wet. By **its** being entirely on the surface you have the opportunity of employing the whole bed quite from the bottom, which **could not be** so well effected if part were buried **in a** trench. **If it** be designed to have

the

the bed in one of the dry kitchen garden quarters, in a rich light foil, make a fhallow trench about fix inches deep, in order to ufe the earth thereof in moulding over the bed, to fave the trouble of bringing it from a diftant part; efpecially where confiderable ranges are intended, and require great quantities of earth ; ufing alfo the earth between the beds, digged down as low as the bottom of the dung, that the whole on each fide of the bed may be cleared fufficiently to admit of fpawning it quite from the bottom. The bed fhould be four or five feet wide, four or five feet high, and in length it may extend from ten to fifty feet or more. If two or more beds are intended, let them be arranged parallel one befide the other, at fix or eight feet diftance, and, if convenient, ranged South and North, that both fides may have equal benefit of the fun's influence, for occafionally

drying

drying the covering of litter more effectually, when rendered wet by exceffive rains.

According to the above directions, mark out the places for the beds, and let the furface of the ground be well cleared from weeds and rubbifh four feet wide, and if a trench is intended, excavate it only about ten inches deep, laying the mould equally to both fides ready for moulding the bed when fpawned.

In the formation of the bed different methods are practifed; but I never found more than one to be good and fuccefsful. Some are made by a layer of dung a foot thick, and a layer of earth, alternately; but beds made entirely of dung are what I recommend; dung and earth together rarely fucceed, notwithftanding this method is recommended by fome eminent writers, who how-

ever

ever appear to have been totally unacquainted with the proper management.

Let the dung when **duly prepared in the heap as** before advifed, be **brought in long and fhort together as it comes to hand :** then having a handy **two-tined fork, &c.** begin to form the foundation **of the bed by** fhaking fome of the longeft dry **litter, evenly** at bottom, forming the **bed at** firft to the full width, and gradually narrowing up- wards, **by** drawing in each fide moderately **and** regularly, generally advancing only a yard or two **in** length, raifing it **by** degrees to a ridge **the full height, as a** guide to the whole ; and continuing it along regularly lengthwife in the fame proportion. Beat the dung firmly in with the fork from time to time as you proceed, and be careful to form both fides of an equal flope, nar- rowing very gradually upwards till they meet

and

and terminate at top in the fharp ridge before mentioned; each end to be alfo pro-
portionally floped. Let the whole be **firmly**
wrought to preferve effectually the requifite
uniformity, and prevent fettling too **con-**
fiderably; for it fhould be three feet, **or**
three feet and a half perpendicular height
when fully fettled. Finifh the **work** by trim-
ming up all fmall dung **on the ground**
around the bed, to the top; **beating the**
whole on both fides **firm and even; fo**
that the bed **now** finifhed **may affume the**
fhape of the roof **of a houfe, both fides form-**
ing fteep flopes, in which **the fpawn is to**
be planted.

In a week or fortnight after the bed is
made **it** will heat violently, and probably
continue fo for a fortnight or three weeks
or more, efpecially if of a confiderable extent,
and muft on no account be fpawned till the
violent

violent heat fubfides and becomes reduced only to a gentle warmth, otherwife the
fpawn will be totally deftroyed and the
whole work to be done over again, and this
is often the caufe of fo many Mufhroom
beds proving barren, the fpawn perifhing
at the firft fetting off. See *Spawning the bed,*
page 36.

When the bed is made, thruft down fome
long fharp pointed fticks, two three or
more, in each bed according to its length,
and by drawing up the fticks two or three
times a week, and feeling the lower end, you
will be able to judge more readily of the
working and ftate of the beds, for the reception of the fpawn.

Let the bed be fully expofed to the open
air, day and night, that its heat may
come on gradually without burning; if

E   exceffive

exceffive rains fhould happen, caft fome dry litter at top, or fpread garden mats, fo as to fhoot off the wet, left it fhould perifh the bed, or occafion it to heat violently and burn; either of which would render it totally ufelefs. Great humidity is a certain enemy to Mufhroom beds, as it foon exterminates the whole fpawny fubftance.

Some perfons indeed make the beds under an airy covered fhed, or barn, or erect a fort of awning of canvas : fome alfo, having confiderable ranges of glafs houfes, make them in thefe departments. I however have always found fuccefs in the open ground, and generally much better than when under any covering.

By way of curiofity and experiment, I have made a bed for Mufhrooms in the fame manner as for cucumbers and melons,
permitting

permitting it to remain till the heat had in a manner quite declined, then put on the frame and placed the spawn on the surface of the dung, and earthed it two inches with light fine loam, covered the whole, half a foot, with dry litter, as also the outside of the bed and frame, defending it with the lights tilted behind, and have succeeded.

I have observed, in the Autumn, in an old melon hot-bed a large quantity of strong spawn overspreading the surface of the dung within the frame, and running considerably through the mould, which was loam. Covering the surface of the earth with dry hay and litter round the outside, and puting on the lights, I have suffered the whole to remain undisturbed till about February, when the Mushrooms began to appear in as great a crop and as fine as

ever

ever were feen. In late cucumber and melon beds, **made** in April and May, for **the** Autumn **crops, I** have, when the heat **of** the **bed** has become very moderate, **placed** fome pieces of fpawn along the edges **about two** inches under the mould, and in Autumn, **have** produced good Mufhrooms, obferving **to cover** the place with a little dry litter.

However **the only** certain method is to **make a** regular bed as before directed.

Spawning the Bed.

IN the work of fpawning the bed, the utmoft precaution muft be obferved, not **to** perform it until the great heat has paffed off, and left only a very gentle warmth; for the fmall tender fpawny-fibres and minute knots

of

of embryo plants would, by one day's great heat, be totally deftroyed. All that is required, is a kindly warmth juft to fet the fpawn in motion, and forward it in fhooting out its tender fibres over the dung and earth. But it muft be remembered, that a bed being fpawned and clofely covered over with the neceffary coat of earth, an inch or two thick, thereby excluding the outward air, and confining the heat within, occafions that heat to be renewed afrefh, and might caufe the bed to burn; fo that you muft be cautious in putting in the fpawn while much heat remains : nor muft the covering of litter be applied too foon after, efpecially in ftrong beds : for thefe require a week, a fortnight, or more, before this is proper to be done.

Be careful therefore in thefe particulars: for on fpawning and covering in at a due degree of warmth, depends the whole fuccefs;

and

and in this you will be regulated according to the working of the bed, as some will be fit to spawn in two, three, or four weeks, others not in less than five or six, according to their length, and the strength of the dung. A bed of fifteen or twenty feet long will be sooner ready for spawning than one of forty or fifty.

After the bed has been made a fortnight or three weeks, examine it frequently by the trying-sticks, which we advised, examining them frequently and you will readily discover the requisite heat and proper state to admit of spawning.

Sometimes in very substantial beds, after they have remained seemingly long enough, and we are doubtful of an increase of heat, we begin spawning on the lower part of the bed first, which part becomes warm before

the

the upper; the heat naturally mounting upwards, and remaining hot longeſt towards
the top; befides, by leaving the upper half
unfpawned and un-earthed, the heat from be-
low if it ſhould prove a little too ſtrong, finds
vent above; but in about a week's time
ſpawn it wholly: the lower part having a
week's advanced growth, will probably fur-
niſh a ſmall gathering ſome days before the
upper half.

However, in general, after having obferved the neceffary precautions juſt given,
take the firſt opportunity to perform the
ſpawning, lofing no time for the bed to ex-
hauſt itfelf ineffectually without being planted.

Let the ſpawn be brought forth in a dry
day, and be careful that it is tolerably dry in
itfelf; proceed to plant it in pretty middling
lumps; not feparating the ſpawn, from the
                                        lumps

lumps of dung in which it is contained; but observing that the large cakes be broken into moderate pieces. Plant the fides of the bed in one or other of the three following methods, *viz.* juft within the dung, earthing over an inch or two thick—on the furface, and then earthing over—or, by firft earthing the bed an inch or two thick, then fpawning the earth, and adding an inch depth more over the whole.

Each method perform as follows.

Spawning in the Dung.

The fpawn being in moderate lumps, is to be depofited juft within the dung, at regular diftances, in rows length-ways beginning the firft within half a foot of the bottom, making fmall apertures by gently raifing the dung a little with one hand, whilft with the other you directly infert the lump, proceeding in
the

the fame manner with the reft, placing them five or fix inches diftance in the row, and the rows about fix or eight inches afunder, a little more or lefs, proportionably either to the *abundance* or *goodnefs* of the fpawn. If your fpawn be in plenty it may be planted clofer, and let the fmall crumbs remaining at laft be laid evenly along upon the top of the bed, which finifhes the article of fpawning.

Then fmooth the fides of the bed with the back of the fpade evenly, for the reception of the cafing of earth, which fhould be an inch or two thick, evenly laid over every part.

Choofe for this purpofe any good, light, rich kitchen-garden earth. If the bed is made in any of the kitchen-ground quarters, you may ufe the adjacent earth on each fide; or, if there is a fhallow trench made, let the

F   excavated

excavated earth be ufed, being careful, firft to break it fine quite down to the bottom of the bed, that no part of it may be loft under ground; then begin the cafing or coat, firft along the bottom, continuing it regularly up the fides of the bed, beating it lightly with the back of the fpade in laying it on, thereby fixing it even and fmooth: thus proceed regularly over each fide, both ends, and the top, fmoothing the whole in a neat manner.

Then place down your long, fharp-pointed flicks, in the fides of the bed, for occafionally trying the internal ftate of the heat, after being clofely earthed over, in order to difcover when to apply the covering of litter, &c.

The covering of litter will be required as foon as you difcover that there is no danger of burning,

burning, which probably may be in a few days, or a week, in moderate beds; in others two or three weeks. This you will readily judge of by the sticks placed in the bed, as above, or according as the weather proves more or less favourable; heavy rains, &c. may oblige you to cover in sooner than you intended, in order to preserve the spawn.

For the purpose of covering, you may provide either clean straw, or long dry horse-stable litter, sufficient to lay about half a foot thick at first, but gradually increased afterwards of due thickness to defend the bed effectually from the air, rain, and inclement weather, and to preserve a low kindly warmth.

As soon as you apprehend all danger from heat to be over, let the bed be finally covered up with the aforementioned dry stable litter, or clean straw, observing to shake the cover-

F 2

ing

ing on lightly with a fork, nearly a foot thick; at firſt, we cover only about half a foot, increaſing it by degrees, and ſometimes only the lower half of the bed, if we are any ways doubtful about the afterheat, ſo gradually advancing upwards till the whole is covered over. It is alſo adviſeable in Winter, and all bad weather, to ſpread large thick garden mats all over the litter or ſtraw, &c. both to ſecure it the better from being diſplaced by the wind, and to ſhoot off the rain before it penetrates too much, ſo as to wet the litter conſiderably, or go through to the bed, which muſt alſo be carefully looked into after exceſſive rains, and if the litter next the bed be wet, to be removed as ſoon as poſſible, **and dry litter applied in its place.**

Spawning

Spawning on the Surface.

IF you have plenty of fpawn this is a
good method to have a forward and plentiful
crop, as the fpawn may be laid pretty clofe
all over the furface, and earthed over two
inches thick, as follows.

Begin the fpawning along the bottom firft,
as in the other method, quite down at
the lower edge of the bed; placing the pieces
of fpawn flat-ways upon the furface of the
dung, befide one another, either clofer or wider
afunder, according to the plenty, proceeding with a regular layer all along, a foot wide,
up the fide of the bed, earthing this over
two inches; then another layer of fpawn
higher up the bed, and earth it as the other,
and fo on till finifhed, laying the fmall
crumbs of fpawn remaining at laft upon the
top of the bed, earthing it over as before directed,

rected, fmoothing the whole with the back of the fpade; and place fome fticks down to difcover the working of the bed and tempe-rature of the heat, as in the former method.

The covering of litter or ftraw muft alfo be applied in due time, when you perceive no appearance of after-heat, obferving the fame precautions and method as already ad-vifed in the preceding article of *Spawning in the Dung*.

Spawning in the Earth.

This is performed by previoufly earthing the bed, and then inferting the pieces of fpawn into the earth, which often proves very fuccefsful.

The bed is firft beat fmooth with the fpade, then earthed all over evenly about two inches
deep;

deep; then, breaking the fpawn into moderate lumps, introduce them into the earth at
fmall diftances all over the bed; and, when
finifhed, add a little fine earth over the whole
near an inch thick, fmoothing it off with the
fpade, as in the other beds; afterwards, obferving the former cautions, let it be littered
in due time with ftraw, or litter, and mats, as
before directed,

General Culture of the Beds; and Produce.

The covering of litter is to remain conftantly
on the beds, days and night, in all weathers,
only be careful to examine it after hard rains,
to remove the wet litter, as before obferved.
During the Winter feafon, in time of fnow
or cold rains, augment the thicknefs of
the covering both of the litter and the mats.

With

With refpect to the produce, the beds begin furnifhing Mufhrooms in a month or fix weeks after fpawning: fometimes indeed it will be two or three months, but there is no great **fuccefs to be** expected when they are long before they yield their firft crop ; a good working bed, if well fpawned and managed, commonly affords plenty in fix or eight weeks, continuing fometimes for three months together, **rifing** in numerous clufters one under another, **covering the furface of the** bed, fome appearing in embryo, fome larger, **and others, at the fame time,** full fized ; **but** thefe laft fhould not remain long enough to become large flaps, becaufe they would prove detrimental to the adjoining fucceffional plants, efpecially when the bed is in **full pro-**duction.

The Autumnal beds, if the fpawn is in perfection, generally produce in a fhorter time

time than thofe made in the middle of
Winter, and Spring beds more freely than
thofe of the hot time of **Summer.**

When it happens that a bed difappoints
our expectation, if, upon examination, the
fpawn appears in life and health, and fmells
well, you are not to difturb it too haftily, **for**
fometimes, after remaining dormant **feveral**
months, a bed will break forth all at once
into confiderable crops. To affift fuch beds
we fometimes, in Winter efpecially, if the
heat appears to be greatly declined, apply a
quantity of moderately warm ftable litter over
the whole, having firft fome dry litter imme-
diately next the bed, then the warm litter a
foot thick over that; which often, by its
kindly warmth, vegetates into life the inac-
tive fpawn.

G   Be

Be very careful to fee that the beds remain fufficiently defended with proper dry litter, never expofing them to the open air, in cold weather efpecially, except juft to gather the produce; or, occafionally, when they have received too much wet, in order to dry the furface for an hour or fo in a fine day; or to remove cafual wet or decayed litter next the bed, till frefh is added in its room; directly covering the whole over again of the proper thicknefs with perfectly dry litter.

If after exceffive rains, the covering receive wet, fo as to penetrate a confiderable way through, let it be as foon as poffible, turned off with a light fork, in a dry time of the day; removing the wet litter next the bed quite away, and directly adding fome dry.

Likewife

Likewife when the litter by long lying on the bed decays, or becomes any way dungy, it fhould be removed and frefh dry litter applied.

In very cold weather, when beds not naturally worn-out, fuddenly decline, it is for want of a proper warmth, which try to recover by applying warm dry litter, as already mentioned.

In very dry hot weather occafionally open the beds then in bearing, and refrefh them with a moderate fprinkling of water, or a moderate fhower, covering them up again.

Of gathering the Mufhrooms.

Though the firft production is fometimes fix or eight weeks or more after fpawning

before

before it appears, at the end of a month
begin to examine the progrefs and .working
of the bed, and if fuccefsful, you will
difcover the running and knotting of the
fpawn abundantly; the Mufhrooms will foon
after begin to advance plenteoufly all over
the bed, when they may be gathered as they
are wanted.

In proceeding to gather them, chufe dry
weather, efpecially during the cold feafons,
and turn off the litter on one fide firft.
Gather thofe above the fize of good middling
round buttons, with a gentle twift of the
hand, head and ftalk together; and be
careful, in their clufters not to difturb the
young fucceffional ones which are advancing
juft within and out of the furface; lay them
gently in a bafket, and fearch quite to the
bottom of the bed; not permitting any to remain

main to become large flaps unlefs fuch **are**
particularly wanted, as fometimes they are.

As foon as you have finifhed gathering,
cover the bed over again directly with the
litter, and if in Winter with mats alfo.

If the bed is in full production it will
probably afford two or three gatherings
weekly, afterward not above once a week
or fortnight, but generally examine it
once a week, as long as it is expected to
bear.

A Mufhroom-bed feldom furnifhes any
abundance after two or three months; it
has often done its beft in fix weeks.

When, however, the bed has totally
ceafed to produce, it will furnifh a fupply
of fpawn for other beds, and the dung will
be

(54)

be excellent manure to wheel on the kitchen
ground. Be careful in pulling it to
pieces, to preſerve the freſh good lumps of
ſpawn, and lay them up dry, as formerly
directed, till they ſhall be wanted for new
beds.

FINIS.

This Day is Published.

THE

Britiſh Fruit-Gardener;

AND

ART OF PRUNING:

COMPRISING,

The moſt approved Methods of RAISING and PLANTING every uſeful FRUIT-TREE and FRUIT-BEARING-SHRUB, whether for Walls, Eſpaliers, Standards, Half-Standards, or Dwarfs:

The true ſucceſsful Practice of PRUNING, TRAINING, GRAFTING, BUDDING, &c. ſo as to render them abundantly fruitful:

AND

Full Directions concerning SOILS, SITUA-TIONS, and EXPOSURES.

By JOHN ABERCROMBIE;

Author of MAWE's *Every Man his own Gardener: or Gardener's Kalendar.*

N. B. Eight Editions have been printed of the *Gardener's Kalendar.* From a diffidence in the Author, it was firſt publiſhed as the pro-duction of " Thomas Mawe, Gardener to his Grace the Duke of Leeds, and other Garde-ners :" though written by Mr. Abercrombie; whoſe name has been added to that of Thomas Mawe, in all the Editions ſince the *firſt.*